ESTABLISHING AN ARSON STRIKE FORCE

Prepared for
Federal Emergency Management Agency
United States Fire Administration

by
International Association of Fire Chiefs, Inc.
EMW-86-C-2080

This report has been prepared for the U.S. Fire Administration, Federal Emergency Management Agency (FEMA) under FEMA Cooperative Agreement number EMW-86-C-2080. All interpretations and opinions expressed are those of the authors and do not necessarily reflect the views of the government or the International Association of Fire Chiefs, Inc.

**Establishing an
Arson Strike Force**

Table of Contents

Establishing an
Arson Strike Force

Section 1

ESTABLISHING AN ARSON STRIKE FORCE

BACKGROUND

How, in a time of tax limitation initiatives, hiring freezes, and budget cutbacks, can a community increase its capability to attack a serious arson problem? With that question in mind, the U.S. Fire Administration commissioned this project. A number of communities had reported to the U.S. Fire Administration that arson strike forces had dramatically improved their success in combatting serious arson problems.

As the name implies, a strike force focuses the capability of a large number of skilled investigators to tackle a major arson or series of arsons. Typically, these investigators come from a number of agencies who agree to share investigative resources when a serious arson problem arises. For example:

The Bureau of Alcohol, Tobacco, and Firearms' National Response Team's (NRT) success in resolving arson cases in many locales around the country. One of the NRT's best known cases occurred in responding, in 1987, to the Dupont Plaza Hotel arson in San Juan, Puerto Rico. Nearly, one hundred died in this arson. Despite the severity of the fire and the complexity of the investigation, the NRT was successful in clearing the case and bringing swift justice to the arsonist.

A strike force approach was used in Ventura County, California to deal with a series of arson fires at citrus processing plants.

In the Detroit, Michigan area state, county, and municipal investigators have used the strike force approach to crack down on a city-wide arson spree that for the past several years has marred Halloween with hundreds of set fires.

In Tennessee, Shelby County and the City of Memphis pooled their resources to break a strip-and-burn auto theft ring that had been responsible for more than 99 such incidents in the Memphis area.

On the surface, the use of a stand-by, special-duty force to handle large or serial arson cases would appear to be a cost-effective approach to managing such cases. But how well does the concept work? And can it be applied in communities that are not now using it?

The answer to how well does the concept work is that their effectiveness under all circumstances cannot be guaranteed. Neither this nor any other study has ever rigorously evaluated the effectivenessof arson strike forces. However, the testimonial evidence from investigators who have participated in strike forces is persuasive: strike forces work and work well. Indeed the Bureau of Alcohol, Tobacco, and Firearms (ATF) hasestablished 15 regional strike forces composed of Federal, state, and local investigators since 1979. In guidelines developed in 1983, Phillip McGuire, then Assistant Director of the ATF, termed this concept, "when properly managed, has proven to be one of our most effective weapons against arson crimes, particularly in major urban center of the country."

So while the performance of such programs have as yet not been rigorously measured, where they have been tried, officials have generally praised their effectiveness. Further proof of this reaction is the fact that many strike forces contacted in this study were set up as temporary teams to investigateone major arson yet proved to have so many advantages that they gained permanent status.

The answer to the second question (Can the concept be used in communities that do not have one?) must be a qualified yes. The concept is transferable to most local situations. However, each community or regional group should carefully tailor their version of the strike force to meet local requirements.

PURPOSE

These guidelines have been drawn up to help communities to analyze their needs and develop their own arson strike forcecapability. They weredeveloped by local investigators with first-hand experience developing and managing arson strike forces.

This guide outlines the essentials for establishing an arson strike force. It begins with a definition of the concept, presents a brief description of how such a unit might respond to an incident, proceeds through the planning steps for single and multi-jurisdictional strike forces, and outlines key elements in their organization and management.

DEFINITION OF AN ARSON STRIKE FORCE

This guide defines an arson strike force as:
. . .a special purpose, short-term mobilization of a team (or teams) of investigators together with allied resources that applies high intensity investigative efforts to a major arson incident or series or incidents.

An arson strike force is an operational body of investigators. It is not a policy-setting body appointed to oversee local arson prevention and control efforts which are commonly called arson task forces.

In general it can be said that the purpose of an arson strike force is to:

- determine the cause and origin of each fire at the request of a participating agency or jurisdiction.
- develop and share information about:
  - arsonists,
  - their methods of operation and allied criminal activities,
  - investigative techniques.

- provide automatic, mutual assistance to other agencies requiring additional resources for thorough and timely investigation of the circumstances surrounding a fire for civil and criminal purposes.

Under this definition an arson strike force is not:

- a new name for an arson task force when that term is used to describe a policy setting/coordination body. Somecommunities refer to their operational investigative unitsarson task force. This can lead tounderstandable confusion about whether an arson task force is an operational unit or a policy setting council. To minimize confusion, the US Fire Administration uses the term strike force when referring to an operational investigative unit. It uses the term arson task force to describe policy setting bodies
- a substitute for satisfactory local investigative procedures and adequate resources,
- a fire-police joint investigative unit.

Figure 1, below, shows the relationships between an arson strike force and other community agencies and organizations concerned with arson control.

**Figure 1.
Arson System Relationships**

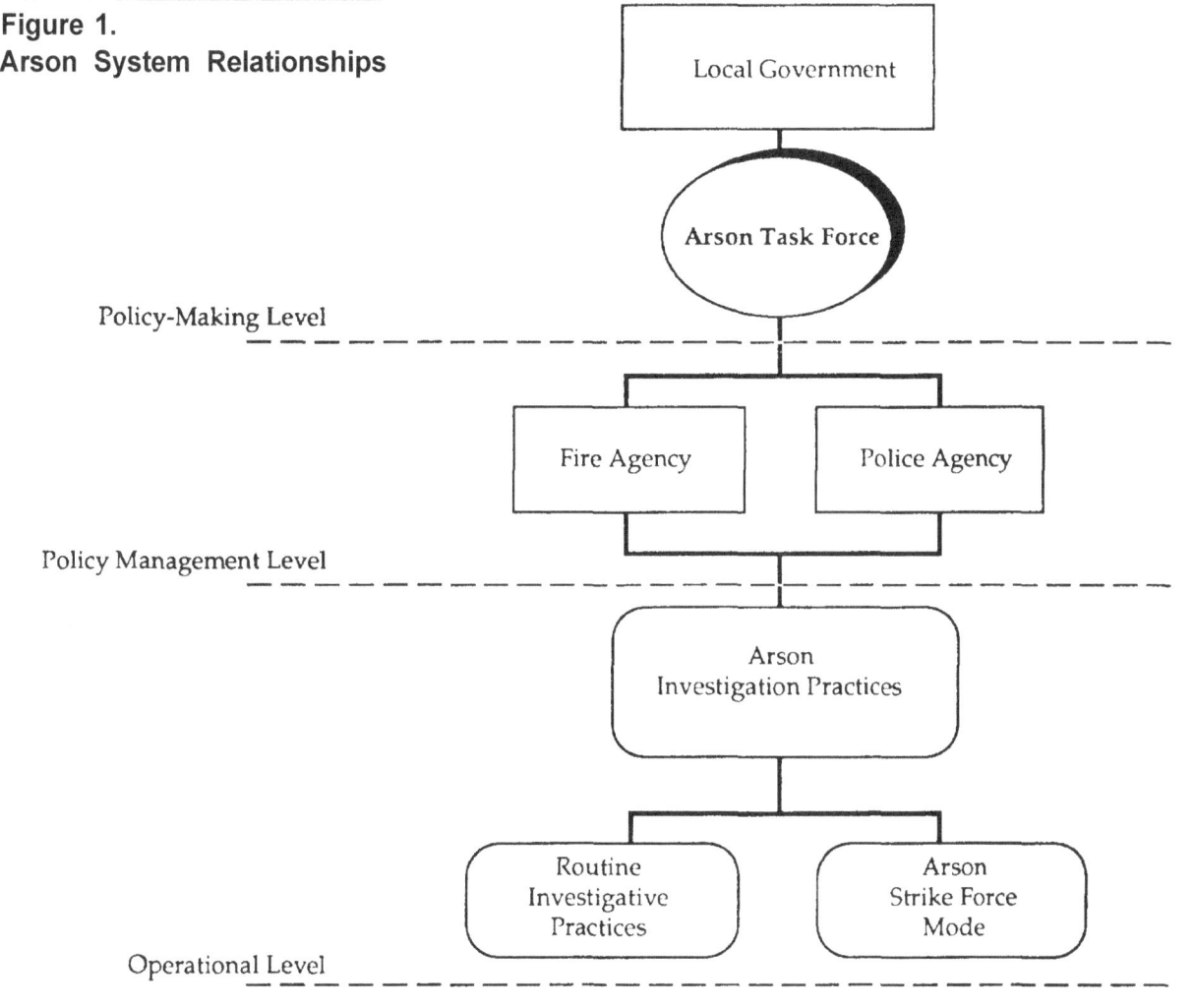

The arson strike force concept typically requires multi-agency planning and coordination. More than one jurisdiction may participate under a reciprocal agreement.

Strike force investigators work such assignments to the exclusion of normal duties. This concentration of skills and resources applies optimum investigative effort at critical junctures in case development with the aim of bringing the case to an early and satisfactory conclusion.

Under the strike force concept, local prosecutors play a pivotal role in force development and operation.

BENEFITS OF AN ARSON STRIKE FORCE

Arson strike forces have sprung up across the country as word has spread of their benefits. Local officials frequently cite a number of benefits, including greater productivity, better inter-agency cooperation, and more arrests coupled with stronger prosecutions. This section discusses three benefits: greater productivity with existing resources, better inter-agency coordination and cooperation, and stronger prosecution.

Greater Productivity with Existing Resources

"We aren't anticipating that arson units will be given much in the way of significant new resources." That is the bottom line consensus that emerged from discussions with local officials. Discussions were held with arson investigators, their unit managers, fire and police officials, and with other governmental leaders. This degree of agreement suggests that significant improvements will have to come from making current resources more productive. One way to do this is cost sharing. And, that in effect is what arson strike forces do for the communities they serve. By bringing together divided resources, overall effectiveness is multiplied. Communities that share expertise reduce the need to maintain redundant resources.

Communities benefit because the team they can field when the need arises should be larger, more capable, and more effective. The team can draw from a larger pool of available investigators. The team offers a second advantage: greater collective knowledge of criminal activity and arson in the area. In turn, as investigators are called out to participate in more arson investigations, the individual investigator's experience base will broaden faster.

In this regard arson investigation is like any other set of professional skills: the more it is practiced in real life situations, the more it is likely that the skills will be sharpened. One of the main problems that investigators face in jurisdictions where few investigations are conducted is skill degradation. This is as true for arson investigation as it is for open heart surgery- it is done best where it is done most. The converse is also true. Fire investigation skills deteriorate quickest in communities where investigators practice these skills infrequently. The same phenomenon holds true for other technical skills. It stands to reason that by exposing investigators to more investigative opportunities than they would get by only working cases in their jurisdiction, that overall capability is improved. By the same token, investigators working in a strike force can practice their special talents. For instance, an investigator skilled in financial analyses might have more opportunities to practice this skill.

Better Inter-Agency Coordination and Cooperation.

Successful fire and arson investigation requires cooperative effort. It calls for a wide range of skills and a critical mass of investigative resources. Strike forces supply these assets in a coordinated manner. Strike forces are set up to simultaneously manage the initial origin and cause investigation, interview witnesses, secure evidence, obtain legal advice, and coordinate with appropriate public and private organizations, including the media.

Exchanging intelligence about known or suspected arsonists is another form of inter-agency coordination. Sharing intelligence about arsonists and their methods across jurisdictional boundaries is vital to stopping the serial arsonist. Interviews with convicted torches have shown that many rely on neighboring jurisdictions not exchanging vital intelligence. By local, state, federal and insurance investigators exchanging information the benefits are shared. By sharing information, arson strike force members expand their collective knowledge. Better intelligence sharing can lead to faster and surer clearances.

Stronger Prosecutions

Prosecutive appeal of the case can be enhanced by the well coordinated efforts of a team of investigators. The first 48 hours following a fire is frequently cited as the window of maximum opportunity for investigators to develop the key leads in a case. While follow-up investigation and preparation for prosecution are also vital phases of an investigation, it is in these first hours where the added capability that a strike force provides can make the crucial difference between making a case and failing to develop it This is particularly true of conducting timely interviews of owners, occupants, witnesses and suspects before memories lapse or alibis strengthen.

SUGGESTED STEPS IN USING THIS GUIDE:

If you are considering establishing an arson strike force in a single jurisdiction:

1. Read the definition (above) a couple of times until you feel you understand it.

2. Skim Section 2, Strike Force Operations, to get an idea of how such a unit might function.

3. Read, in order, Sections 3, 4, and 6 comparing the suggested steps with your situation. Make notes as you go along.

4. Re-read Section 2 and think about how the elements you've just described in your notes could combine in a strike force operation.

5. If at this point you think the strike force concept is applicable in your jurisdiction, begin step-by-step implementation with Section 3.

If you are considering establishing an arson strike force in a multi-jurisdictional setting:

1. Read the definition (above) a couple of times until you feel you understand it.

2. Skim Section 2, Strike Force Operations, to get an idea of how such a unit might function

3. Read, in order, Sections 4, 5, and 6 comparing the suggested steps with your situation. Make notes as you go along.

4. Re-read Section 2 and think about how the elements you've just described in your notes could combine in a strike force operation.

5. If at this point you think the strike force concept is applicable in your are, begin step-by-step implementation with Section 4.

Section 2

STRIKE FORCE OPERATIONS

The following description and accompanying flow chart are intended as an aid to users of this guide. Their purpose is to help the reader visualize how an arson strike force would function. It is a simple, generic outline of an investigation in which a strike force is activated.

A "GENERIC" STRIKE FORCE OPERATION

1. A suspected arson fire has occurred or is in progress.

2. The local arson investigator believes that circumstances may justify activation of the arson strike force (Alternately, the local fire officer in charge may determine the need to call out the strike force).

3. The investigator confers with the fire officer-in charge (OIC) at the scene regarding use of the strike force.

4. The investigator (or 01C) contacts and confers with the arson strike force commander. The investigator briefs the commander on the circumstances and the reasons why the strike force should be activated A frequent reason given is that the investigative requirements are beyond the skills **or** available resources of the requesting agency.

5. The strike force commander activates necessary elements of the strike force (especially those who will work the scene). Roles and responsibilities are assigned according to an incident command system (ICS).

6. A strike force meeting is held to review initial findings and to assess the situation. Decisions will be made regarding whether and how to proceed with the investigation.

7. Specific investigative and administrative assignments are made and a reassessment meeting is scheduled. Typically investigators would be assigned to one of the following roles: origin and cause, evidence control, photographing and diagramming the scene, or interviewing. Others would be assigned support responsibilities as required. These might include planning, logistics, and media relations.

8. Investigative teams begin working their assignments. When strike force members complete their initial assignments (including paperwork), they

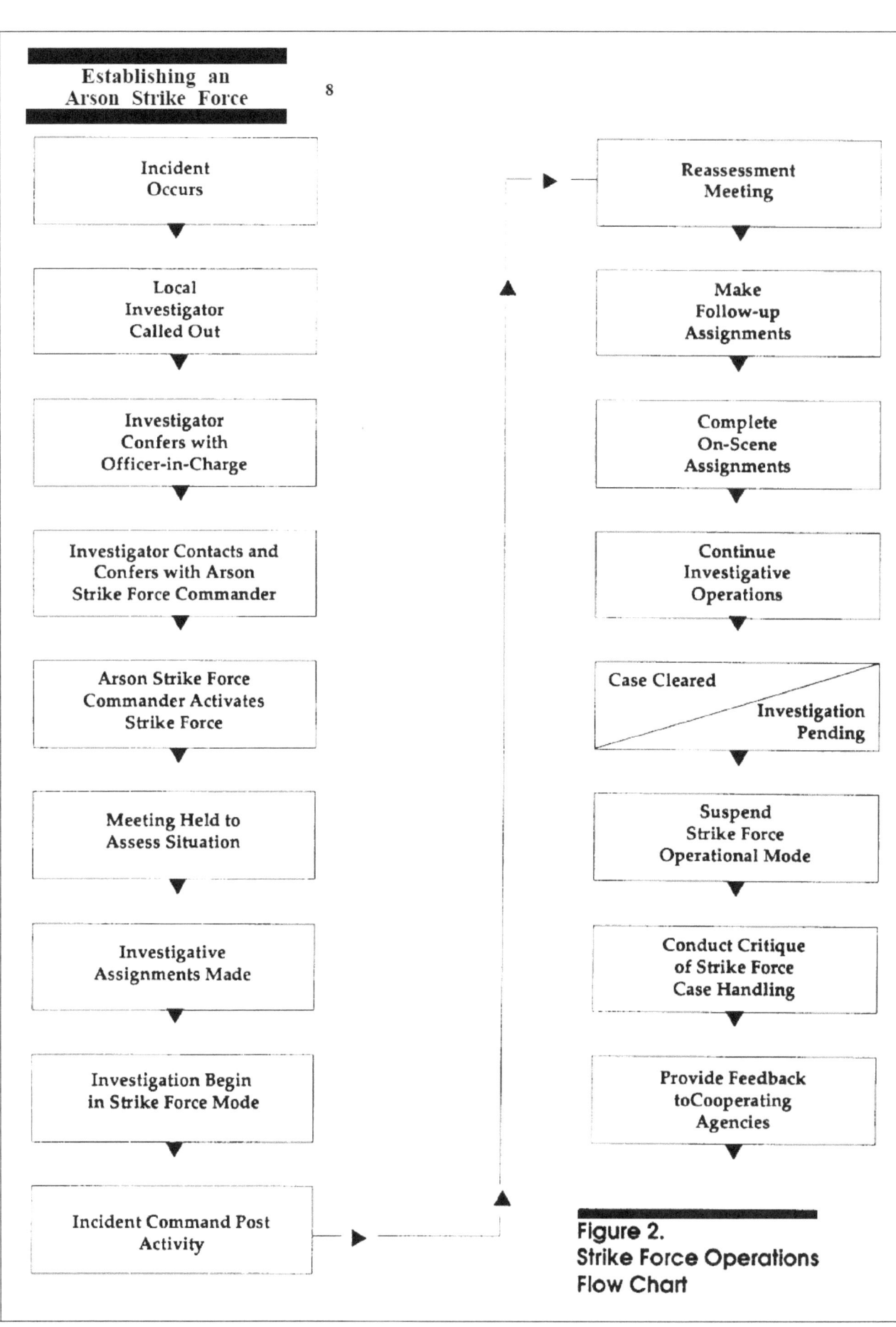

Establishing an Arson Strike Force

Incident Occurs

Local Investigator Called Out

Investigator Confers with Officer-in-Charge

Investigator Contacts and Confers with Arson Strike Force Commander

Arson Strike Force Commander Activates Strike Force

Meeting Held to Assess Situation

Investigative Assignments Made

Investigation Begin in Strike Force Mode

Incident Command Post Activity

Reassessment Meeting

Make Follow-up Assignments

Complete On-Scene Assignments

Continue Investigative Operations

Case Cleared / Investigation Pending

Suspend Strike Force Operational Mode

Conduct Critique of Strike Force Case Handling

Provide Feedback toCooperating Agencies

Figure 2.
Strike Force Operations Flow Chart

may be assigned other duties or be released back to their own agencies.

9. If the incident warrants, the strike force will set up an incident command post at or near the scene.

10. A m-assessment meeting is held a day or two after the incident to review progress and set the course of the follow-up investigation.

11. Investigators receive follow-up assignments. Some members/teams may be de-activated while others are called out.

12. On-scene activities are wrapped up.

13. Investigative operations continue, but the focus of the activity shifts away from the scene. In an arson-for-profit case, the focus of investigation may shift to the "paper chase," collecting and analyzing documentary evidence.

14. With sufficient investigative progress, the case can be cleared; otherwise it will have to be referred to "pending" status at some point. Either way, investigation cannot continue indefinitely. Periodic (daily, if necessary) meetings will m-assess case status and direction.

15. At some point, active strike force operations will be suspended. This may be done all at once or through a gradual scaling-down of effort. Provision should be made for disposition of case materials.

16. A post-investigation critique of the case and strike force performance is held to review results and recommend changes/improvements.

17. Cooperating agencies and jurisdictions are informed of progress and disposition of the case so that they can see the effect of their participation.

Section 3

CONSIDERATIONS IN DEVELOPING A SINGLE JURISDICTION'S ARSON STRIKE FORCE

The following descriptions outline suggested steps in planning an arson strike force in a single jurisdiction. Each item corresponds to an element of the flow chart in Figure 3. These steps dove-tail with those described in Section 5, Organizational Development. Thus, to progress through the elements in developing an arson strike force in a single jurisdiction requires that the reader proceed from the end of this section directly to Section 5.

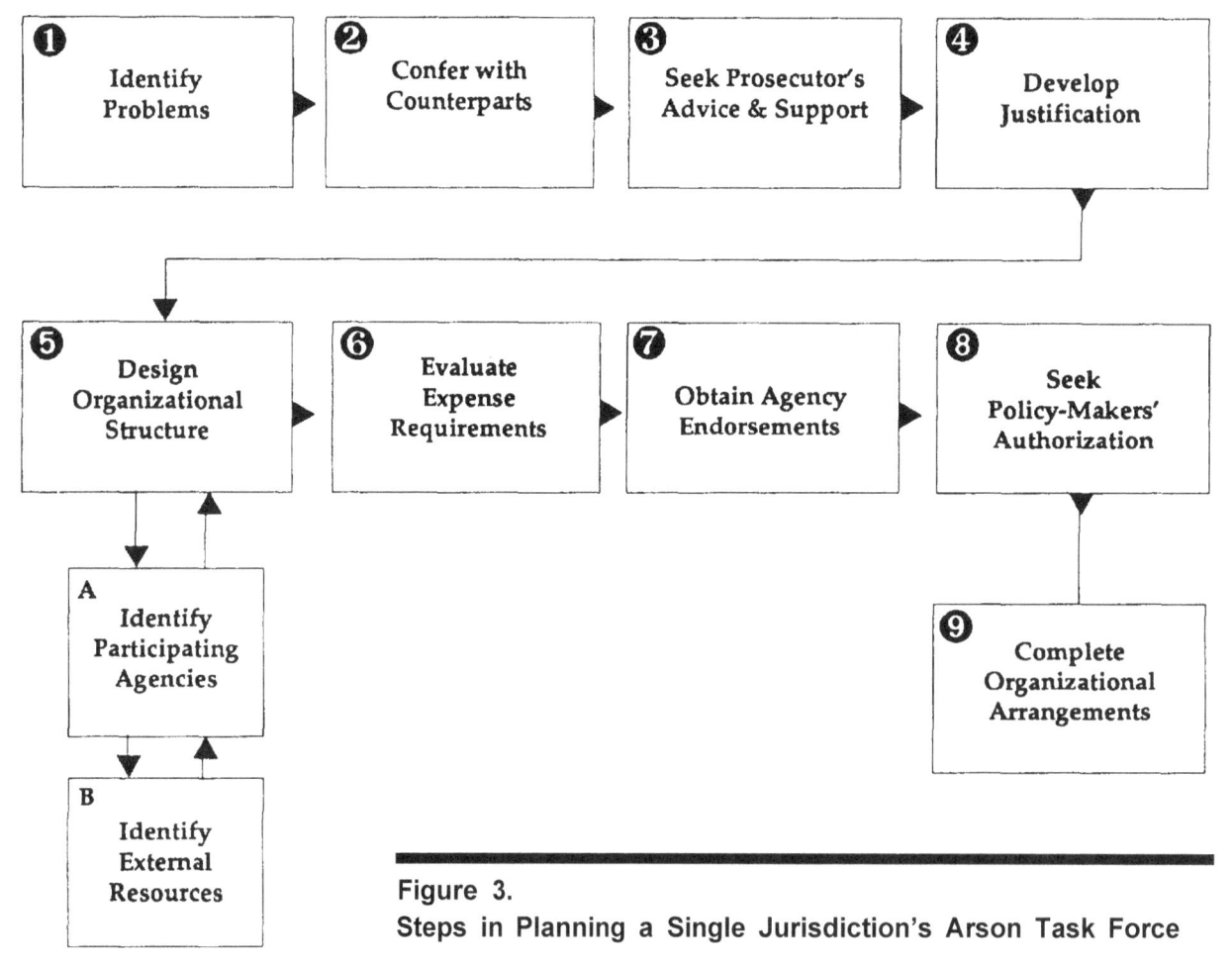

Figure 3.
Steps in Planning a Single Jurisdiction's Arson Task Force

STEPS IN PLANNING A SINGLE JURISDICTION ARSON STRIKE FORCE:

1. Identify Problems

This should include but not necessarily be limited to consideration of the following:

- the number and types of arson documented as occurring in the jurisdiction
- retrospective examination (perhaps over the previous five years) of the conduct and outcome of the investigations of major and/or serial fires (include all incidents meeting these criteria, not just known arsons)
- adequacy of existing investigative mechanisms in dealing with large or serial fires: (i.e., Would a high intensity effort have led to a more satisfactory outcome?)
- benefits to be gained from establishing the strike force.

2. Confer with Counterparts in Other Agencies

Fire service executives should meet with their law enforcement counterparts (or vice versa) to discuss problems identified in step one and potential applicability of the strike forceoption. Law enforcement officials may be aware of other instances in which the strike force concept has proved effective in dealing with homicide, narcotics, or organized crime activity. This may predispose them to the benefits of the concept when applied to arson investigation.

3. Seek Prosecutor's Advice and Participation

Ultimately, the success or failure of the strike force approach will depend on the involvement of prosecutors. Therefore, the District Attorney or State's Attorney should be consulted as early as is practical in the development of a strike force. The prosecutor's support for and input to the strike force formation will ease its development. Specific understandings should be sought from the outset on the strike force's role, composition, and procedures, and the role of the prosecuting attorney's staff in the strike force. Working out these understandings in advance will later aid cases that the force develops to be effectively prosecuted. And convictions, not just clearances are the hallmark of successful arson strike forces.

4. Develop Justification

Justification for the strike force approach is likely to come from:

- data documenting that major arson incidents or serial arsons would be better resolved through a concentrated effort
- staff and/or budget cutbacks that result in a need to address an arson caseload with diminished resources.

Justification might include noting that the strike force constitutes a standing reserve of investigative capability at little additional cost. This would be an especially potent argument in favor of a strike force if the problem identification phase shows that shortagesof personnel and resources have seriously hindered previous investigationsof major and/or serial fires. If appropriate, point out the prudence of establishing the strike force before a major arson requires establishing it on a crash basis.

5. Design Organizational Structure

Designing a general structure for the strike force will require at least three steps. These are:

A. Identifying participating agencies and their roles

B. Identifying applicable external resources (e.g.: state investigative agencies, Internal Revenue Service, postal inspectors, sources of heavy equipment, etc.).

C. Defining the incident command system to be followed.

6. Evaluate Expense Requirements

Strike force activities should not require significant increases in expenditures because they rely primarily on the use of existing resources. Typically, participating agencies will cover the costs of their personnel and resources supplied to the strike force. There will, however, be some costs that cannot be handled in this way. Principles for sharing these costs will have to be developed. They should be simple and fair, and agreeable to all participating agencies.

Some agreements stipulate a time limit for no-cost investigative assistance. For example, the Sierra Front Inter-agency Investigation Association provides that the first 48 hours following assignment will beat no cost to the requesting agency. Time over that amount are subject to reimbursement decisions on a case-bycase basis.

7. Seek Agency Endorsements:

Participating agencies should formally endorse the planned strike force and state their commitment to its support.

8. Seek Policy Makers' Authorization

Inter-agency strike force arrangements in single jurisdictions may require authorization from elected and appointed policy makers before a formal organization can be developed.

9. Complete Organizational Planning

Once policy makers give the green light, the next step is to complete the detailed planning for how the strike force will be organized and managed. Specifics of organizational planning and development (chain of command, personnel selection, etc.) are described in Section 5.

Section 4

CONSIDERATIONS IN DEVELOPING A MULTI-JURISDICTIONAL ARSON STRIKE FORCE

The following descriptions outline suggested steps in planning a multi-jurisdiction arson strike force. Each item corresponds to an element of the flow chart in Figure 4. These steps lead into those described in Section 5, Organizational Development.

In settings where some form of arson-related, multi-jurisdictional investigative organization already exists, the order of steps may differ from those described below. For example, problems will need to be identified by conferring with representatives from other jurisdictions. Letters of support may not be as necessary. An existing agreement governing the multi-jurisdictional investigative body may provide the legal framework for a multi-jurisdictional arson strike force.

STEPS IN PLANNING A MULTI-JURISDICTION ARSON STRIKE FORCE:

1. Identify Problems

This should include but not necessarily be limited to the following considerations:

- document the number and types of arson occurring among the jurisdictions
- review (last three to five years) the conduct and outcome of the investigations of major arsons and/or serial fires (review all fire incidents for patterns, not just known arsons)
- analyze the adequacy of existing investigative mechanisms in dealing with large or serial incidents (i.e., Would the high intensity investigative capability provided by strike force operations have led to a more satisfactory outcome?).

2. Confer with Counterparts in Other Jurisdictions

It will be necessary to identify other jurisdictions that should be involved in any potential strike force. Factors that tend to influence the makeup of a multi-jurisdictional force include politics, personalities, and previous operational working relationships. In reviewing the list of candidate jurisdictions, consider whether or not an adequate depth and breadth of investigative skills will be found among that set of jurisdictions.

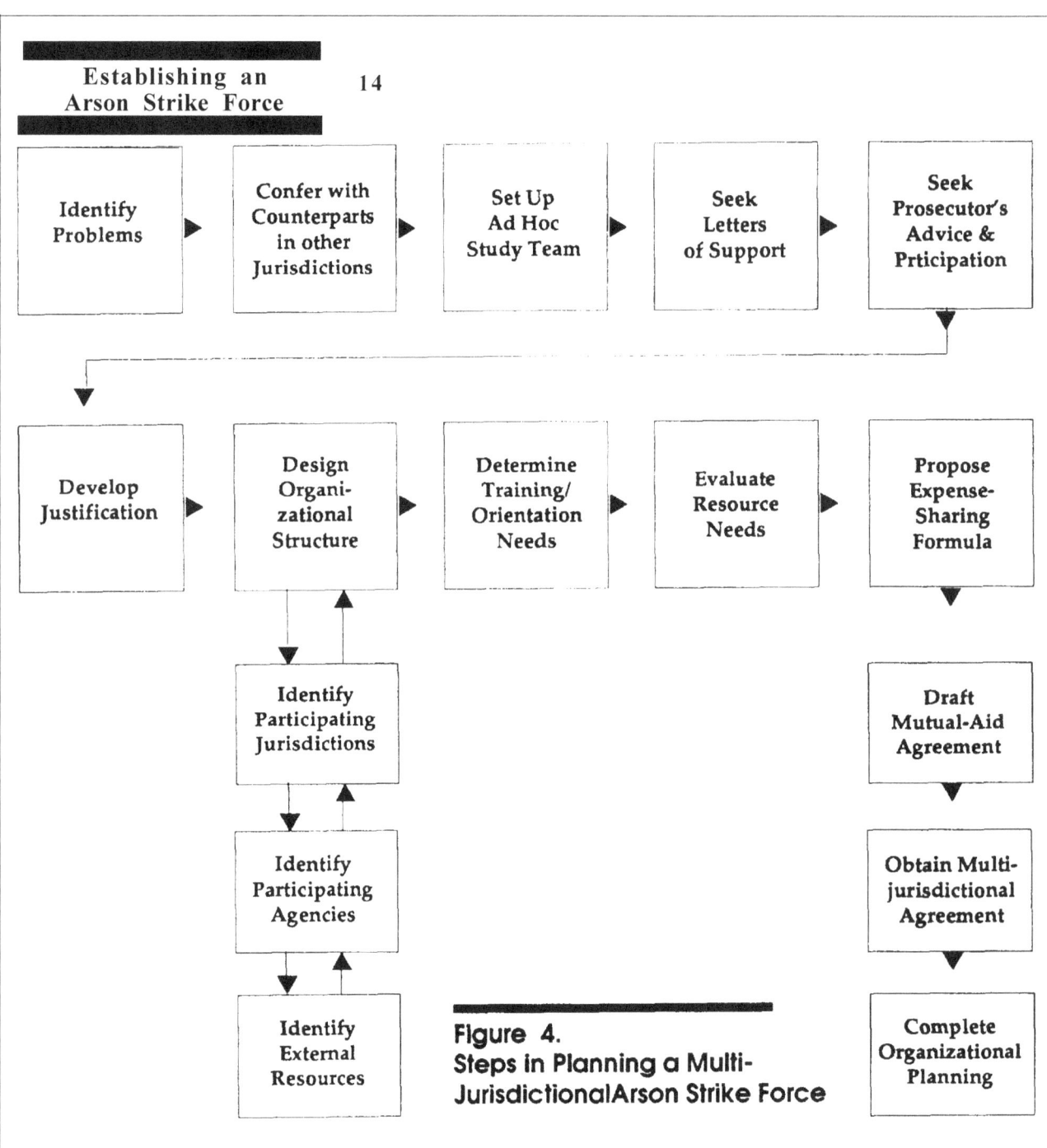

**Figure 4.
Steps in Planning a Multi-
JurisdictionalArson Strike Force**

Consider a multi-jurisdictional approach when any of the following conditions exist:

- arson incident patterns or modus operandi suggest a multi-jurisdictional problem exists
- opportunities exist to improve multi-jurisdictional cooperation
- two or more jurisdictions could benefit from sharing resources
- fire and arson investigators anticipate significant benefits could be derived from closer working relationships.

3. Set up Ad Hoc Study Team

Establish an inter-jurisdictional study team to compile and review arson-related information from all of the jurisdictions involved. Special planning issues facing multi-jurisdictional strike forces that are likely to need resolution are those dealing with political/geographical concerns, training, funding, and personnel matters.

4. Seek letters of Support

If the results of the ad hoc study effort indicate that a multi-jurisdictional approach is both needed and viable, have potential participating agencies declare (in writing, if possible) their support for the concept.

5. Seek Prosecutor's Advice and Participation

Ultimately, the success or failure of the strike force approach will depend on the involvement of prosecutors. It is vital that the District Attorney or State's Attorney be consulted as early as is practical in the development of a strike force. The prosecutor's support for the strike force can influence whether the force enjoys the necessary political support. Moreover, since the prosecuting attorney's jurisdiction may extend over all the communities planning to participate on the force, it may be desirable to organize under the auspices of the prosecuting attorney. Specific understandings should be sought from the prosecutor on the strike force's role, composition, and procedures.

Pay particular attention to potential problems in the prosecution of cases developed by a multi-jurisdictional investigative team. For example, by working with prosecutors the strike team can develop ground rulesabout how to assign investigators in a case to minimize the need to call them as witnesses in later courtroom proceedings.

6. Develop Justification

Justification for the strike force approach can be derived from three main themes:

- statistical data indicating a significant incidence of large or serial arsons whose resolution demands the application of a concentrated effort
- staffand/or budget cutbacks that result need to address an arson caseload with diminished resources
- a finding by the ad hoc study team (or some other qualified body) that either (1) there exists an arson problem that is beyond the investigative capabilities of individual jurisdictions or (2) that by pooling resources joint investigative efforts would enhance investigative efficiency and effectiveness..

The justification should note that the strike force constitutes a standing reserve of investigative capability available to all participating jurisdictions at little additional cost. This will be an especially potent argument if the ad hoc study team concludes that shortages of personnel and resources have seriously hindered previous investigations.

7. Design Organizational Structure

Designing a structure for the strike force will require at least four steps:
- Identifying participating jurisdictions
- Identifying participating agencies and their roles
- Identifying external resources (e.g., laboratory and other forensic services, state fire marshal, Bureau of Alcohol Tobacco and Firearms, state forestry department, Internal Revenue Service, postal inspectors, sources of heavy equipment, etc.)
- Identifying management authority and administrative responsibilities.

8. Determine Training/Orientation Needs

Multi-jurisdiction strike forces will probably need minimum standards for training and experience (acceptable to member jurisdictions and agencies) for personnel. Provision will also have to be made for orienting personnel to strike force rules and regulations.

9. Evaluate Resource Needs

Some form of resource sharing arrangement will be necessary for effective performance. Identify anticipated needs and achieve some arrangement for meeting them.

10. Evaluate Funding Needs and Expense-Sharing Formulas

Strike force activities should not require significant increases in expenditure. Principles for cost sharing should be simple and fair, and must be agreed to by all participating agencies and jurisdictions. When the strike force is being planned, some formula for cost allocation should be set, particularly when participating jurisdictions are of substantially different size or where shared costs (such as special equipment purchases) are involved. Some possible bases for cost allocation are: population, number of incidents, property values, area, or some combination of these. Dividing up costs equally among all jurisdictions is often a reasonable option, especially when the amounts are minimal.

11. Draft Mutual Aid Agreement

A written agreement should be developed to guide all parties and minimize misunderstandings. Written agreements force participants to think through the objectives of the strike force and the ways and means to achieving them.

In a multi-jurisdiction system, a written agreement that sets out the strike force's organizational structure and procedures is a standard practice. It may incorporate training and orientation needs and cost-sharing formulas, Some strike forces have taken advantage of the fact that their parent jurisdictions are organized under existing provisions of a Joint Powers Agreement. Others have been organized under provisions of a state master mutual aid agreement or a statutory provision governing automatic aid.

12. Obtain Multi-Jurisdictional Agreement

Policy makers in all participating jurisdictions of a multi-jurisdictional agreement should approve the arrangements anticipated in the mutual aid pact.

13. Complete Organizational Planning

Specifics of organizational planning and development (Chain of Command, personnel selection, etc.) are described below in Section 5.

Section 5

ORGANIZATIONAL DEVELOPMENT

Once the need for a strike force has been demonstrated, a general structure designed, and approval for implementation received, it is time to proceed to the specifics of strike force development.

All agencies and jurisdictions participating in the strike force must be given a sense that their involvement directly benefits them and that their participation is crucial to the overall success of the strike force. This is the first order of business. Therefore, each should be encouraged to play as large a role as possible consistent with its capabilities.

because much of the strike force's existence will be in a standby mode the following guidelines are suggested:

- Operations should conform as closely as possible to the standard operating procedures (SOPS) of the participating agencies/jurisdictions.

- Where substantial differences between SOPS exist, pay special attention to developing procedures and conducting training that will aid the strike force to perform effectively.

- Write down procedures that spell out the purpose, objectives, duties and responsibilities, request and activation procedures, membership requirements, arrest policies, property/evidence handling, documentation requirements, training objectives, and safety and disciplinary rules.

In structuring the organization, the following considerations will have to be addressed.

ESTABLISH CHAIN OF COMMAND

Establishing a clear chain of command will require careful attention. Local precedent may prove a problem or a help. If, for instance, some inter-agency or multi-jurisdictional force already exists, it may provide a useful model for the strike force. On the other hand, the chain of command for another kind of joint activity may not fit the requirements of the arson strike force. Other factors that will need to be considered when developing a chain of command include: jurisdictional differences (statutory vs. elected responsibilities), the presence and nature of mutual aid pacts, communications requirements, labor and personnel issues, legal issues (e.g. the use of deadly force, peace officer status, arrest powers etc.), and the role of the prosecutor.

Incident Command System (ICS)

How well a major case is managed can determine its outcome. Poor management can doom the case. Moreover, mis-management can threaten the future of the strike force. One management system that minimizes these potential problems and that fire and police officials are increasingly relying on is known as the Incident Command System (ICS).

The Incident Command System has been recognized nationally for its structured approach to managing complex incidents. ICS provides a management structure and system for conducting on site, multidisciplinary operations. It was originally designed in the 1970s to manage wildfires. But since then it has proved so successful for a variety of fire, law enforcement, and emergency management applications that it hasbecome widely recognized as the management system of choice for multi-agency and multi-jurisdictional field operations. It is particularly appropriate to consider as a management tool for arson strike force operations because ICS is gaining such wide use among fire and law enforcement agencies.

ICS is ideal for rapidly activating a management structure tailored for each specific incident. ICS recognizes that while the nature and severity of incidents vary, managing the public safety response to them requires some common management functions. The major functions include command, operations, planning, logistics, and finance.

To manage these functions an incident commander assumes overall control. The incident commander appoints other individuals to take responsibility for each of the four functional areas as needed. In turn, if the workload increases, that individual will further delegate responsibilities to others.

ICS is designed to be modular: to grow or shrink to match incident requirements. Functions that are not needed, as is often the case with finance, are not activated. As the nature of the incident changes, an individual's assignment may change as the nature of the incident evolves. For example, at the outset of a major fire, the arson strike force members may report to a fire commander in charge of suppression operations. Once suppression activities are concluded, the lead investigator may take over as the operations section chief and later may become the incident commander.

ICS operations are designed around other sound management practices, including: management-by-objectives, restricting supervisory span-of-control to no more than five to seven subordinates, and structured information recording and sharing. How these ICS functions and principles could be applied to an arson strike force investigation of a major arson-for-profit incident are illustrated by Figure 5. and described below.

Incident Command

The Incident Commander has overall management responsibility for the incident. Responsibilities include public information, media relations, safety, and inter-agency and inter-jurisdictional liaison. Typically, these functions are delegated to one or more officials who serve as command staff to the Incident Commander.

Operations Section

This section in headed by an Operations Section Chief or equivalent title, who is responsible for the overall tactical management of the investigation. If the incident is a major wildfire, many other sub-units such as air operations may need to be added to this section. The Operations section is responsible for all phases of the on-scene and follow-up investigation. Responsibilities include security, fire scene processing, interviewing/interrogations, case management, follow-up investigations, and any special operational requirements.

Planning Section

The planning section supports the incident commander and operations section by collecting and analyzing all data regarding the incident. ICS planning is based on management by objectives. The planning function includes developing basic operational plans including written objectives, priorities, and strategies so that all involved know and contribute to achieving them. If the length of incident requires it, the planning function prepares one action plan per shift. The planning unit assesses the status of the investigation and reports them together with options for future activity to the incident commander. The planning unit chief is responsible for preparing and conducting all planning and coordination meetings, records management, case documentation, evidence custody and coordination, analyzing intelligence (tips, file data etc.), and case critique data.

Figure 5.
ICS applied to an arson
strike force investigation

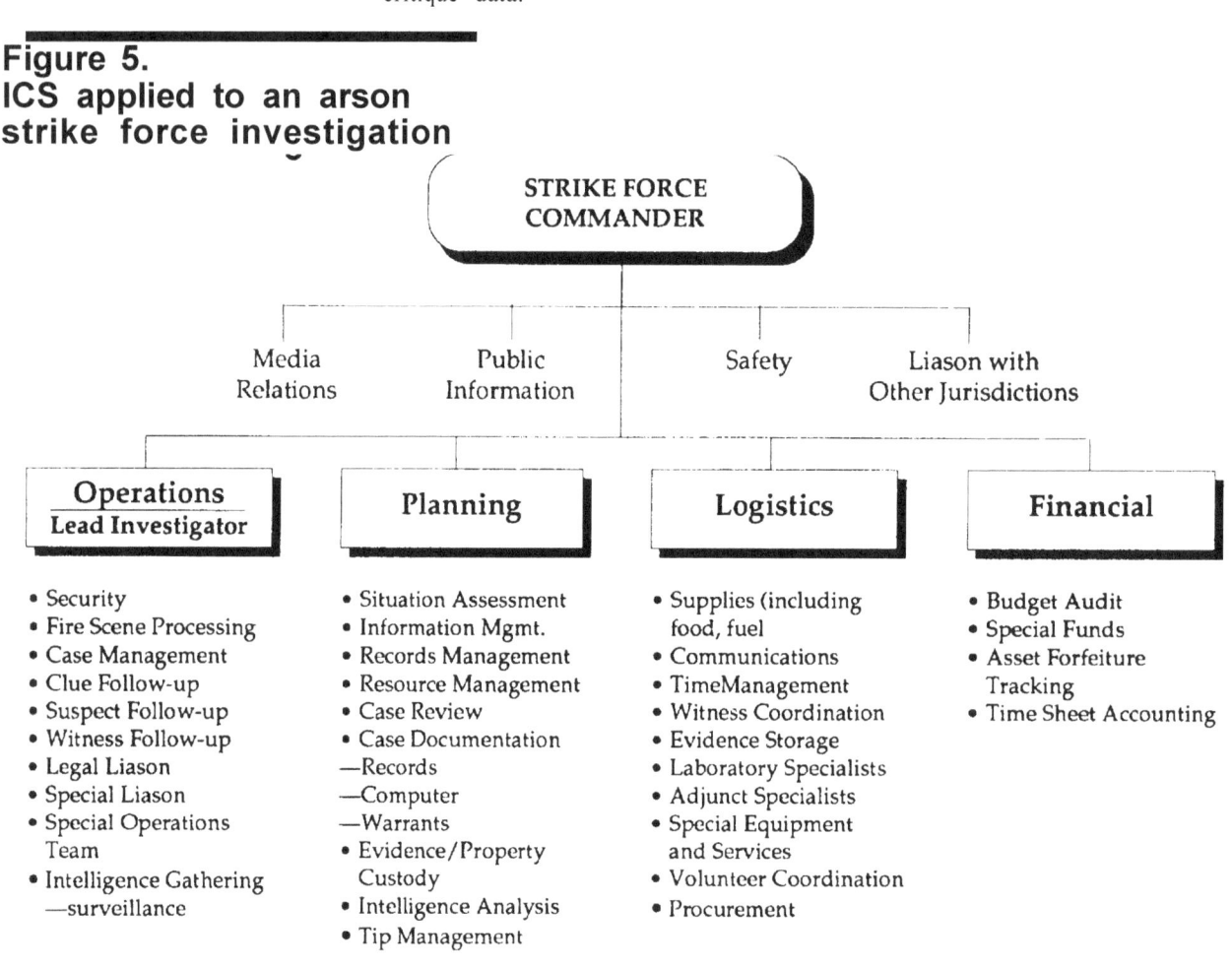

STRIKE FORCE COMMANDER

Media Relations   Public Information   Safety   Liason with Other Jurisdictions

Operations **Lead Investigator**	Planning	Logistics	Financial
• Security • Fire Scene Processing • Case Management • Clue Follow-up • Suspect Follow-up • Witness Follow-up • Legal Liason • Special Liason • Special Operations Team • Intelligence Gathering —surveillance	• Situation Assessment • Information Mgmt. • Records Management • Resource Management • Case Review • Case Documentation —Records —Computer —Warrants • Evidence/Property Custody • Intelligence Analysis • Tip Management	• Supplies (including food, fuel • Communications • TimeManagement • Witness Coordination • Evidence Storage • Laboratory Specialists • Adjunct Specialists • Special Equipment and Services • Volunteer Coordination • Procurement	• Budget Audit • Special Funds • Asset Forfeiture Tracking • Time Sheet Accounting

An excellent source for managing the large scale investigative efforts can be found in Incident Command System publications, particularly ICS 120-1, Operational System Description (See Appendix A. for further information).

Logistics

Responsible for providing the support resources based on the action plan to accomplish the stated incident control objectives. These duties might include obtaining all supplies such as food, fuel, lighting and marshalling other resources as needed such as specialists, volunteers, facilities, and communications equipment.

Financial

The financial function is responsible for documenting the expenses incurred in the investigation, obtaining necessary purchasing authorization, and preparing the paperwork. Financial concerns may not figure prominently in most investigations. However, more lengthy investigations may raise the need to track overtime, track assets for seizure and forfeiture in RICO cases, disburse special funds, and conduct other financial matters as assigned.

SET CRITERIA FOR PERSONNEL SELECTION

When determining who is to be a member of the arson strike force, the following criteria may be applied:

1. Commitment to the team approach; without this commitment the candidate's long term value to the strike force is questionable

2. Membership in a duly organized fire department, police department, or other investigative agency.

3. Ability to meet minimum standards for training and experience. Develop these standards in consultation with all member agencies.

IDENTIFY STRIKE FORCE TEAM MEMBERS

It may take hard work to ensure that agencies assign people with sufficient qualifications to meet task force needs. Some jurisdictions have handled this sensitive issue by forming a credentials committee. The credentials committee asks jurisdictions to submit the qualifications of nominees for consideration. Independent panels reduce the potential that membership is given or denied arbitrarily

Recognize from the outset that there may be problems with the makeup of the strike force. Some agencies may resist committing high calibre staff to the strike force when this could take these people away from their regular duties for extended periods. In a multi-jurisdictional setting, some participating jurisdictions and/or agencies may lack sufficient qualified personnel to allow full participation in the force. However, in the interest of providing for maximum participation, it may be necessary to permit special provisions to accommodate special needs for a limited time.

As important as the number of investigators will be their skill mix. Therefore, it will be necessary to have an inventory of available skills among strike force members. This will allow the force commander to activate the number and mix of personnel needed at each phase of the investigation. Among the skills that should be sought in the members of a strike force or available to them on call are the following:

- Cause and Origin (certified if at all possible)
- Crime scene analysis, photography, diagramming
- Interview/Interrogation
- Prosecution
- Case management
- Computer Operations (arson information management systems)
- Financial Analysis
- Fire Insurance
- Polygraph
- Incident Command
- Link Analysis
  -Investigative Pattern Analysis Chart (IPAC)
- -VIA Visual Investigative Analysis (VIA)
- White Collar Crime Analysis (Financial Paper Chase, Accounting)
- Intelligence skills
  -toll record analysis based on dialed number recorders
  -surveillance
- Media Relations.

DEVELOP/ DETERMINE CALL-OUT PROCEDURES

Define the circumstances and procedures for strike force activation. How and by whom will the strike force be activated? What criteria will be used to decide whether the strike force ought to be activated (e.g.: for fatalities, suspicious losses above a certain dollar value, etc.)? What will be the method for alerting strike force members in the various participating agencies (and jurisdictions) that the force is being activated? Will authorization to use certain personnel have to be sought at the time of activation or can it be granted in advance? Who, in each member's home organization, will have to be notified?

Make advance provisions regarding those who will participate in initial investigation and those who will be consulted during later stages of investigations. Clearly, strike force activation should not be an all-or-nothing proposition. Rarely, will structural fires require fewer than three or more than twelve investigators.

In serial arson case, where possible, assign the same origin and cause team to each additional fire suspected of being part of the series.

ADDRESS ADMINISTRATIVE REQUIREMENTS

The strike force will require administrative support upon activation. There will also be certain on-going administrative and management activities that will need to be maintained when the strike force is inactive. Specific requirements are listed in Section 6.

DEVELOP PUBLIC RELATIONS PROGRAM

Designate one spokesperson for on scene and follow-up coverage of strike force activity. Work with all other agencies and officials to restrict releases to the media to this single source. If the incident is a major news item plan on regular, non-sensitive briefings for news media. Give everyone involved good press. Give out no news that compromises on-going investigations. A proactive public relations programcan payoff during investigationsby bringing in more tips and holding down unauthorized leaks and misinformation. Sound public relations can also maintain support for the continued operation of the strike force.

For a good review of principles to follow in dealing with media during a major investigation, the reader is referred to "Media Relations", chapter 9 of the National Institute of Justice's Multi-Agency Investigative Team Manual (See Appendix A. Resources and References for additional information).

Section 6

MANAGEMENT AND ADMINISTRATION

STEPS IN DEVELOPING MANAGEMENT AND ADMINISTRATION

This section describes ten steps to manage and administer an arson strike force. Figure 6. illustrates these steps which are outlined in more detail on the next Page.

1. Assign Administrative Responsibility

Two sets of administrative needs will have to be attended to in support of the strike force. The first set addresses the needs for on-going, sound administration of the strike force. These administrative duties need not require much effort, but they do need regular attention. Examples include maintaining up to date rosters, training records, and similar paperwork. Assure that "housekeeping" details are attended to even when the unit is inactive.

The second set deals with the support structure needed when the strike force is called out. These needs include anticipating the consequences of extended investigations, insuring the availability of needed resources, computer support, and media relations. Both sets of needs will need on-going efforts to keep the strike force functioning smoothly.

The strike force's administrative arrangements will have to include clear assignment of responsibility to ensure the continued maintenance of the following:

- standard operating procedures for the strike force
- chain of command
- membership recruitment, retention and composition of investigative team(s)
- staff support
- call-out procedures
- duty rosters and scheduling
- communications
- forms and reporting formats
- civil liability coverage
- logistical support
- arson information management systems
- training and education.

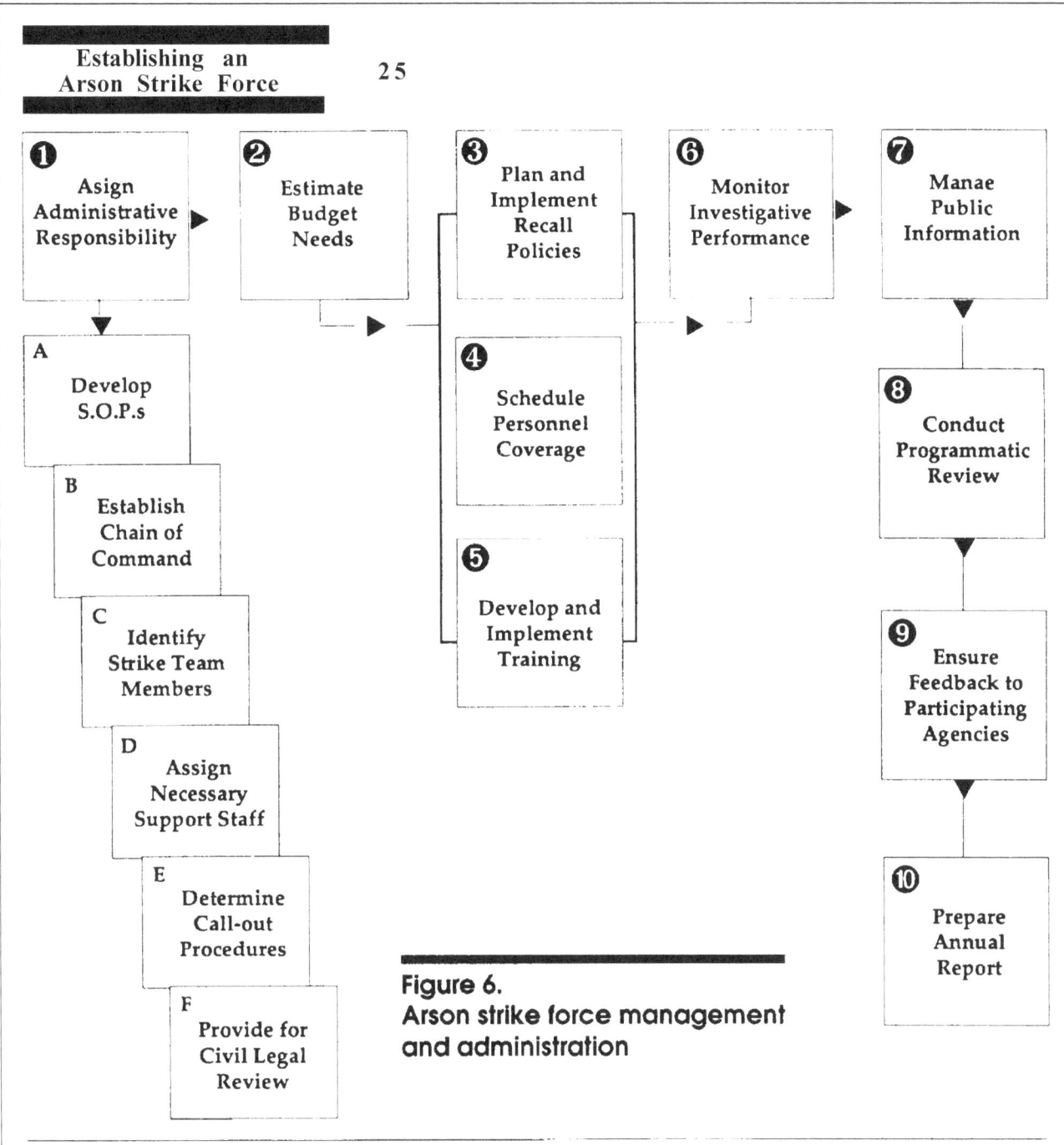

Figure 6.
Arson strike force management and administration

It would be wise to review management and administrative arrangements after the strike force has been activated two or three times.

2. Estimate Budget Needs

Most stand-by arson strike forces operate without requiring funds to be earmarked for their operation. As a rule of thumb, each agency is usually expected to pick up the cost for its personnel. Under this arrangement the jurisdiction where the investigation is required remains responsible for those non-personnel expenses associated with the incident. In some jurisdictions, operational costs can be recovered through mutual aid funds.

In developing a budget for the strike force, wherever possible, re-program funds to allow it to operate within existing budgets. With or without a line item budget, consideration should be given to contingency planning for extraordinary operational expenses. Such expenses might be for special surveillance equipment or extraordinary overtime costs. The mechanism for obtaining funds for specific expenses under special circumstances should be arranged in advance. Then, if this provision is ever needed, an established procedurecan be followed rather than requiring the task force to scramble to make arrangements on the spot.

3. Establish Activation and Recall Policies

Specific, agreed-upon policies will be required for the activation of investigative teams. In managing call-out policies strike force administrators may need to pay attention to two areas in particular: night time call-out and deactivation procedures.

Night time call-out procedures
Call-out procedures need to balance the need for quick response with the reality that fireground conditions may restrict all strike force resources from being effectively utilized on arrival. As a practical matter there are a number of factors that can delay investigators from immediately working the fire scene. These include the increasing need to obtain search warrants, delays due to the need to remove safety hazards, and the need for ventilation and limited overhaul. Therefore, those managing the call-out process need to consider what resources need to be phased in when. Thus, while interviewing and observation of the fireground and crowd can commence on arrival, cause and origin may be postponed for hours. Careful size-up of fireground requirements and constraints can limit amount of time that investigators spend idling on scene.

Deactivation procedures
If the fire is determined to be accidental will the strike force remain at the disposal of the requesting jurisdiction until the initial investigation is completed or will the strike force complete the investigation.

4. Schedule Personnel Coverage

Even though this is a standby unit, some arrangement will have to be made to allow quick activation of critical elements of the strike force. Once the strike force is activated, the scheduling responsibilities turn from call-out coverage to relief coverage to ensure that strike force members aren't overstressed by fatigue. Strike force managers have found this can become a particular challenge during lengthy investigations.

Rotation to Equalize Workload
To equalize workload no investigator on the call-out roster should routinely have two call-outs in a row. An exception to this rule of thumb is when a team member has a fire in his own jurisdiction following a response to another jurisdiction and no one else is able to cover. Another exception is in a serial arson when the cause and origin team needs to kept intact for investigative continuity.

Assignment Flexibility

Attention should be paid to assign the right individuals to the right task. This is especially true for the more sensitive aspects of a case. For example, an investigator with special rapport with a volunteer fire department might be selected to pursue follow-up interviews with volunteersover an otherwise equally qualified investigator. The point here is that both in terms of the basic agreement establishing the strike force and in terms of its operations, there should be sufficient latitude to permit the strike force leader to exercise discretion in making assignments and other decisions.

5. Develop and implement Training

Arson strike force members often cite training for all members as one of their toughest problems. They usually add that this is true for both basic and advanced training. Frequently, the members of such strike forces have other obligations which limit their availability for training outside their normal scope of duties. Scheduling difficulties and lack of funds may further restrict training opportunities.

This presents challenges if the strike force adopts the goal of getting every member fully cross-trained in all aspects of fire investigation and follow-up arson investigation. Thus, fire investigators may find it difficult to find the time to achieveand maintain statestandards Peace Officers Standards Training [POST] or its equivalent). Similarly, criminal investigators may find it difficult to complete certification for the National Fire Protection Association professional standards for investigators. The same obstacles make it difficult to send a member away for advanced training at the National Fire Academy or Bureau of Alcohol, Tobacco, and Firearms Arson-for-Profit Investigative Course at Glynco, Georgia.

Despite these difficulties, finding the time and resources to ensure adequate training is available to members of the strike force is cited frequently as the glue that holds multi-jurisdictional strike forces together.

As a minimum training should be arranged to instruct all members of the strike force and key people in contributing agencies of the strike force's concept of operations and procedural guidelines. Close monitoring of investigative performance as noted below will also help define investigative needs.

6. Monitor Investigative Performance

Strike force managers have the responsibility to monitor investigative performance. After action critiques of investigations can point out problems not only in investigative performance, but also in case management, and overall strike force operations. Left uncorrected minor shortcomings can compromise subsequent investigations. Candid evaluations of performance can head off most serious problems developing. To cite results from the a review of serial murder investigations, five case management problems stood out as the most frequently encountered:

- lack of overall administration
- lack of overall coordination, ongoing case review and analyses, in-

formation management
- too many non-investigative responsibilities for investigators,
- undefined investigative priorities
- poor interaction with the news media Consult the state's attorney/ district attorney as part of this review of case management.

While these findings come from the field of homicide investigations, they indicate the serious issues that can be surfaced and dealt with by closely monitoring investigative performance. To complete the monitoring process, managers will have to develop and coordinate improvements indicated by case critiques.

Some of the matters that should be evaluated as part of the critique process include:

- fire scene processing
- investigative operations, techniques, and tactics
- evidence collection, storage, and custody
- personnel selection, assignment, supervision and control, any use of volunteers, consultants and other professionals
- overall management of the strike force
- records management
  -arson information management systems
  -computer operations
- communications within the strike force and with other agencies (fire, police, prosecutors, forensic specialists etc.)
- support operations (planning, logistics, and finance)
- security (all phases)
- safety
- training
- interaction with prosecution, court personnel.
- legal issues
- public relations/interactions with witnesses, public figures, and media.

7. Manage Public Information

Actively cultivate the local media's awareness of the strike force, its objectives, and benefits. Both the public at-large and policy makers are influenced by what they see and hear in local media. Some strike forces have developed specially marked uniforms and vehicles to promote their visibility to the public, especially through the media. The New York City Fire Department's "Red Cap" program is one example of this technique. Publicizing strike team operations, their clearances and convictions can pay-off by deterring would-be arsonists.

As part of the public relations program, seek to strengthen interpersonal relations between agency and jurisdictional heads in order to minimize potential friction points. Joint news conferences and annual awards programscan play important roles in an active campaign to foster support for the strike force among decision makers.

8. Conduct Programmatic Review

Conduct periodic reviews of the strike force program. This review should be conducted by persons independent of the strike force, and should be followed by consultation with the leader of the strike force. It may be possible to arrange for a courtesy audit from a law enforcement agency's inspection unit. The Commission on Accreditation for Law Enforcement Agencies publishes standards that can aid the development of an appropriate set of review guidelines. These standards may be obtained through local law enforcement agencies.

9. Ensure Feedback to Participating Agencies

Agencies and jurisdictions taking part in the strike force should be made aware of progress resulting from their participation. Communications through both briefings and meetings are needed but they must be well planned and directed.

Recording a briefing is one way to reach your intended audience with the greatest impact. Tape record or videotape update briefings for fire fighters/police officers who have contributed to the development of the case. These briefings, of course need only contain non-sensitive information. The briefing need not compromise any aspect of the investigation. The briefing can stress the contributions of firefighters and others to the case development. Either a videotape or tape recorded briefing can be produced without consuming an inordinate amount of time. Since future investigations will also depend on the cooperation and contributions of these officers, any efforts to acknowledge their contributions to current investigations are likely to pay future dividends.

10. Prepare Annual Report

An annual report covering force activities, case status, and results of programmatic review will provide a basis for assessing strike force performance and utility. The report can help maintain awareness among policy makers and the public that the strike force is worthy of their continued support. Even if the force is not activated for some time, take advantage of every opportunity to remind all agencies involved and the public at large of the benefits of the arson strike force.

CONCLUSION: FIVE CHALLENGES

The cost of maintaining full investigative resources for a major arson investigation is a commitment of resources that few communities have made. The best realistic alternative is for jurisdictions to maintain basic resources to meet their normal investigative workload and to initiate a strike force concept that mobilizes additional resources in a timely, cost-effective manner. This solution applies equally well in urban, suburban and rural settings. The result is that investigations of major incidents are concluded faster and with a greater probability of clearance and conviction. Strike force operations can make the individual investigator more competent by exposing him or her to more and varied fire scenes. Benefits also accrue to the individual agencies and jurisdictions in the forms of cost avoidance, improved inter-agency coordination, and better clearance and conviction rates.

Experience has shown that those seeking to establish and maintain an arson strike force commonly encounter fivechallenges. How well these challenges are met often determines the long-term success of the strike force. To summarize they are:

- selling the arson strike force concept to policy makers

- obtaining multi-agency commitment

- defining an effective incident management system

- following sound management practices

- maintaining commitment of investigators, their sponsoring agencies, and elected policy makers.

This guide has been developed to aid others willing to accept these challenges and willing to spend the time necessary to meet them. There wards of those who have met these challenges include safer communities and the pride of professional accomplishment that comes from teamwork towards the common objective of arson control.

Appendix

RESOURCES AND REFERENCES

For more information on arson strike forces please consult the following resources:

The US Fire Administration maintains an Arson Resource Center as part of the National Emergency Training Center Learning Resource Center (LRC). Write or phone the LRC to request additional information on arson strike force organization and management

Learning Resource Center
National Emergency Training Center.
16825 South Seton Avenue.
Emmitsburg. MD 21727.
(301) 447-2787 or (800) 638-1821

The National Institute of Justice sponsors the National Criminal Justice Reference Service (NCJRS). NCJRS is the Nation's primary reference and referral clearinghouse for criminal justice information. Contact NCJRS for further information on arson control and allied law enforcement operational techniques.

National Criminal Justice Reference Service.
P.O. Box 6000.
Rockville. MD 20850.
(800) 851-3420 or (301)251-5500

Additional Written References

U.S. Fire Administration. Arson Resource Directory Washington. D.C.: Federal Emergency Management Agency. 1988.
This resource guide contains listing for both arson task forces and strike forces. It is an invaluable guide for those seeking first-hand advice from others active in arson prevention and control.

Brooks. Pierce et al.. Multi-Agency Investigative Team Manual. Washington. D.C.: National Institute of Justice. 1988.
This manual offers additional guidance in managing complex. multi-agency investigations. Chapter 9. Media Relations deals with media management. Written for multi-agency homicide units. its points are apply as well to arson strike force operations.

Three Incident Command Systems publications can be obtained from the International Fire Service Training Association. IFSTA offers several ICS publications including the ICS 120-1 Operational System Description manual, a package of Field Operations Guides, and a complete set of position descriptions for command, operations, planning, logistics, and finance. Contact them for current prices. Write or phone:

Fire Protection Publications
Oklahoma State University
Stillwater, OK 74078
(405) 624-5723

www.ingramcontent.com/pod-product-compliance
Lightning Source LLC
Chambersburg PA
CBHW081242170526

45165CB00009B/3155